Jorge Alfredo González Pérez
Eunice Jeaques Velinay

Agro-productive behavior of two varieties of Couve in Cuvelai

Jorge Alfredo González Pérez
Eunice Jeaques Velinay

Agro-productive behavior of two varieties of Couve in Cuvelai

Behavior of 2 varieties of Couve in the edaphoclimatic conditions of the locality of Mukolongodjo, municipality of Cuvelai

ScienciaScripts

Cover image: www.ingimage.com

This book is a translation from the original published under ISBN 978-3-639-68271-7.

Publisher:
Sciencia Scripts
is a trademark of
Dodo Books Indian Ocean Ltd. and OmniScriptum S.R.L publishing group

120 High Road, East Finchley, London, N2 9ED, United Kingdom
Str. Armeneasca 28/1, office 1, Chisinau MD-2012, Republic of Moldova, Europe
Managing Directors: Ieva Konstantinova, Victoria Ursu
info@omniscriptum.com

Printed at: see last page
ISBN: 978-620-8-56981-5

THANKS

✓ I'd like to start by thanking Almighty God for giving me the strength to carry on. To everyone who, in one way or another, made their skills and efforts available to make this stage a reality.

✓ I would like to thank my mother, who has always supported me in my academic training.

✓ I would particularly like to thank my husband, Jorgelino Afonso dos Santos, for his support and encouragement throughout this journey.

✓ In particular, I would like to express my deepest gratitude to my advisor, MSc. Jorge Alfredo González Pérez, for his reading of the work and the methodological suggestions he made with a view to better restructuring it.

✓ I would like to thank the I.P.O. and all the teaching staff for the teaching conditions they provided during my training, my family in general, my colleagues and friends who all helped to make this moment in my life possible.

My deepest thanks to everyone

DEDICATORY

- ✓ This work is dedicated to my eternal parents Jeaques Moisés Eduardo and Colmicil Ndamona (in memory), for being the first people to encourage me to take the first steps towards an educational institution.
- ✓ To my husband, Jorgelino Afonso dos Santos, for his patience, understanding and support.
- ✓ To my son Jorgeveny Peandju Velinay dos Santos, for being my second blessing.

SUMMARY

This research was carried out to evaluate two varieties of cabbage under the soil and climate conditions in Mukolongodjo. The experiment was conducted at the R. L Cantinho do Chicote cooperative, in the municipality of Cuvelai, and the varieties analyzed by treatment are: **T1** Cabbage (Brassica Olerácea L. *var. troncuda o português*), **T2** Cabbage (Brassica Olerácea L. *var. manteiga* I -1811 *(B)*). Growth variables assessed: Plant height (cm), number of leaves (U), pseudo-stem diameter (cm) and leaf width and length (cm). Yield variables and their components: Fresh Leaf Mass (g). The best results for the growth and leaf weight variables were obtained in treatment T1 (Cabbage (Brassica Olerácea L. *var. troncuda o português*)) for the local soil and climate conditions. The economic results were positive for both treatments

Keywords: soil and climate, leaf, fertilization, conditions.

Table of contents.

I. INTRODUCTION

The world's population and technological capacity to preserve life and food has grown progressively. Over the last 200 years, growth has been exponential, which means that the world's population is doubling every 40 years, according to Cassinda (2016), who also points out that tropical countries, especially those in Africa, the vast majority of which belong to the third world, suffer the most from the effects of overpopulation, malnutrition, social inequalities, environmental deterioration and are the most vulnerable to the challenges of the future.

Agriculture and its production represent the basis and sustenance for most of the world's peoples. The increase in the quality of life on the planet depends on its scientific advances and technological progress, which is why meeting the food needs of a rapidly growing population is one of the essential challenges of the 21st century. In the future, these needs would have to be guaranteed fundamentally by increasing the productivity of agricultural crops, especially those that constitute the food base in each region FAO (2012) and cited by Gaudioso (2021).

The total area of Angola, according to the FAO (Aqua STAT), is around 124.6 million hectares and the cultivated area in 2007 was no more than 3.6 million hectares, or just 2.88 % of the country's total area. For a country in which agricultural activity is the main source of employment (agriculture and livestock represent the main economic activity of more than 65 % of the Angolan population, according to MINADER (2003)), this figure is clearly very low, showing that farming is still at a very early stage. Gaudioso (2021).

Angola is one of the countries with the greatest potential for development on the African continent. Having been affected by the international crisis of 2009, one of the most dynamic sectors for the growth of the Angolan economy is Agriculture, with a real growth rate of 29 %, and the agricultural sector is vital for the economy and society, according to the General State Budget (OGE, 2011).

However, according to Neto *et al.* (2006), oil dominates the country's economic situation and is the main enemy of agriculture, with the country continuing to rely heavily on imports, particularly of food products. This has led to the transformation of agricultural and rural Angola into an Angola dependent on a mono-product - oil, which has brought with it the impoverishment of the majority of the population. He also stresses the difficulty of revitalizing Angola's agricultural sector and developing the rural world.

In a report presented by the Agricultural Sector Development Plan (PDSA) and the National Statistics Institute (INE, 2015), agriculture is the third largest sector in Angola's GDP, although less than 30% of arable land is cultivated and agricultural productivity and crop yields are still very low compared to other countries in sub-Saharan Africa.

Authors such as Sambeny, Alves, Calegar, Tollini, Dias *et al.* (2011), agree that agriculture in Angola is predominantly a family activity for thousands of small subsistence farmers, who plant an average of 1.4 ha per family, with the planted area increasing slightly every year.

Production has been carried out in the main rainfed planting season between September and April. The second planting season takes place mainly in low, wetlands and is carried out from June to August.

Vegetables are considered to be one of the world's most important foodstuffs due to their importance in the diet, their high nutritional content of both vitamins and minerals, and the fact that they are a consumption option for the world's inhabitants. According to Castro, Díaz, Pérez, Rodrigues and Gómez (2010), the world's population has grown progressively. In the last 200 years, growth has been exponential, which means that it doubles every 40 years, so our technological and scientific capacity must be geared towards preserving life and feeding ourselves. The world's great challenge in the coming years will be to produce food to satisfy the growing demand of the population that will inhabit the planet.

In the countries of the African continent, agriculture is at the heart of the challenges faced by these countries, especially the third world countries, because it is the source of a large part of Africa's food and other raw materials, and is home to the most unequal population, which has a significant influence on stabilizing the environment and reducing the productive capacity of the soil (García, 2004).

For every human being, vegetables represent the only source of nutritious sustenance to produce energy, regulate bodily functions, nourish themselves and live. From the economic and social point of view, they are important because they are a source of food and employment, because of the demand for food in all walks of life, and because of their high value and industrialization in local, regional and national markets.

In terms of food, they are considered important to the human diet as they provide the human body with vitamins, minerals, carbohydrates and fiber, which are essential for the normal development of the individual and the prevention of many diseases (Bettiol, 2004).

Among vegetables, cabbage (Bróssica olerácea L.) is gradually becoming a very important product for farms, and yields need to be increased. This crop is gaining more and more acceptance as a result of the change in consumer tastes, which is evolving into a preference for using units that are not very large, but which produce more (Saavedra, 2013).

Cabbage is a plant that is difficult to describe, as the different varieties are very different in morphological terms. It can be considered a herbaceous plant, but there are some varieties that are woody in the stem area, and it can be considered a biennial plant, but it sometimes has perennial inclinations, and its life cycle can extend for more than two years. At the end of the Middle Ages, cabbage was the most widely used vegetable in Europe; curly or smooth varieties were already eaten in Greece in the 4th century BC. The ancient Romans already referred to kale as liana, and it is considered to be the most ancestral form of today's kale. Today we can differentiate it to the length of the stem and the characteristics of the leaves (Almeida, 2014).

It is a food rich in nutrients and low in calories. It is rich in vitamin A and C. It has calcium, beta carotene and a high anthocyanin content (Cassinda, 2016). Organic vegetable production, in the case of cabbage (Bróssica Oleraceae L), has grown significantly over years, causing growers or farmers to change the way they plant or manage it.

It is also important for human nutrition, has been recognized for thousands of years and is an excellent source of energy for a healthy diet (Saavedra, 2013).

According to Cartea, Velasco, Obrégon, Padilla and De Haro (2008), vegetables are often eaten raw. But there are situations in which straining is necessary or even preferred. In this case, the nutrient content and capacity these vegetables can be altered. With positive and negative consequences such as the improvement of naturally occurring compound activity, the formation of new compounds with nutritional activity and the loss of naturally occurring nutrients.

Kale is rich in iron and chlorophyll, making it an excellent choice for preventing and treating anemia, and it helps to detoxify the body, preventing the absorption of chemical substances from food and fluid retention. Kale is also rich in folic acid, an essential vitamin before and during pregnancy, which helps the development of bone marrow in babies during gestation (Zamberlam, 2001).

The high content of vitamin C and sulphur in cabbage eliminates toxins (free radicals and uric acid), which are the main causes of arthritis, skin diseases, rheumatism and gout (Meihua Clinic, 2014).

Every year, the province of Cunene imports a large quantity of this vegetable from other provinces in Angola and the Republic of Namibia, spending a lot of money on it, and a considerable amount of resources, which increase as consumption and prices rise (MINADER, 2005).

For this reason, introducing and evaluating different varieties of vegetables over time, and cabbage in particular, would be a good step forward in the fight against hunger, hence the formulation of the **scientific problem** in this research: "Which variety of cabbage has the best agro-productive performance in the soil and climatic conditions of Mukolongodjo".

That's why we came up with the following **hypothesis:** "If we can determine which variety of cabbage performs best, then we can increase the yields and production of this vegetable in the town of Mukolongodjo."

The following **general objective was** proposed: "To evaluate the performance of two cabbage varieties in the soil and climate conditions of Mukolongodjo".

Specific objectives

1. To evaluate the main agro-productive characteristics of cabbage (*Brássica olerácea* L) in Mukolongodjo.
2. Select the variety with the best results.
3. Analyze the economic costs of growing cabbage.

II. LITERATURE REVIEW

2.1 General characteristics of the agricultural sector

One of the fundamental characteristics of production systems is their effect on improving or maintaining the properties of the soil and environment, and the production or productivity of the systems. This is a particularly important aspect to consider when developing mechanisms for assessing sustainability and production systems, as it would guarantee a comprehensive assessment of the management-environment-production system (Ríos, Munoz, Pérez, Franco and Arango, 2000).

According to Márquez (2012), in many countries, the agrarian sector is characterized by its heterogeneity in the use of production systems, whether it is a traditional subsistence-oriented peasant society, whose production is carried out with local inputs, conservation practices and family labour, peasant societies in transition with a mixed agriculture partially oriented towards the market, supported by a mixture of traditional and modern technology and reciprocal labour relations, and on the other hand, a modern capitalist rural society with specialized and mechanized capital-intensive production, and dependent on the international market.

As the United Nations Conference on Trade and Development points out, there are two ways of increasing agricultural production: increasing arable land and increasing crop yields. However, the possibilities for increasing arable land are limited, as is the availability of water for agriculture. In this sense, a lithographic proof from the FAO (2012) states that 80% of the increase in production needed can come from an increase in yields and only 20% from the expansion of arable land.

The agricultural and livestock sector is a key economic sector for any economy. It is classified as an economically complex sector, which contains extensive technical and scientific knowledge that has been accumulated, passed on and enriched from generation to generation. The production of food is the first condition of life for direct producers, and of any production in general, so the livestock sector is a decisively important sector in material production and for the economy (Muñoz, Germani, Aguilar, Jacinto, Sarmiento *et al.*, 2015).

Agricultural productivity is one of the most important agronomic aspects in the sector, as it is increasingly necessary to increase production and, at the same time, not expand the planted area, so efficient management of the available resources is essential for maintaining and improving the population's living conditions (Paterniani, 2001 and Carvalho, Ferreira, Silva Teixeira, Oliveira, Carvalho and Santos, 2016).

2.2 Production systems

A production system is a combination of several subsystems: crops, defined at the level of homogeneously farmed plots, with the same technologies and sowing successions (in this sense, several agricultural systems are distinguished within a production process), livestock, defined at the level of animal herds, the processing of agricultural products (cereal processing, cheese making, etc.), and non-agricultural activities (Apollin and Eberhart, 1999). The self-subsistence system is known as family production since they are the ones who make the most of their productive resources (Palácios and Barrientos, 2014).

Agricultural production systems are facing major challenges due to globalization, and it is very important that these systems improve their management in order to adapt to the various changes in the environment and maintain a sustainable level of competitiveness (Aguilera *et al.*, 2003).

Ecological farming systems are considered to be capable of restoring and conserving soil, water and biodiversity in general. It is therefore necessary to compare them with other systems and production approaches, taking sustainability indicators into account (Astier, Masera and Galván, 2008).

The selection of production systems depends on the interests of each producer, according to the relative scarcity of available resources in order maximize earnings per hectare (Dufumier, 1990), generating interactions between subsistence farming, cash crop production and other means of diversifying earnings.

The need to characterize production systems in contexts of rural poverty in Latin America is due to the great diversity of biophysical and socio-economic conditions, since characterization is fundamental to designing bio-diverse, resilient and socially just strategies for the consumption of humanity (Altieri and Nicholls, 2013).

The standardization and weighting of indicators, based on quantitative procedures, makes it possible to compare properties and analyse the multiple dimensions (environmental, economic and social) of the sustainability of production systems (Sarandón *et al.*, 2006).

2.2.1 Farmers' knowledge of their production systems

There are multiple forms of knowledge in communities that can be recovered for incorporation into the design natural resource management programs (Ortega, 2009), who pointed out that farmers know their crops, production environments and socio-economic conditions well. In many cases, they can express the reasons that drive them to use certain agronomic practices or make certain decisions.

This knowledge, documented by numerous experts in the social sciences and biology, covers the types of soil and production environments in which they work (Kamangira, 1997), their crops and varieties (Sperling *et al.* 1993), the insects and pests they confront (Bentley and Rodríguez, 2001) and the soil and water management practices they use (Lamers and Fiel, 1995). Understanding this knowledge is a fundamental step towards establishing a dialog between farmers and scientists, and a key point reference that farmers use when making decisions and communicating with each other (Lambert, 2010).

Therefore, if scientists want to contribute to farmers' well-being, they need to understand their knowledge, provide them with new information, develop appropriate technologies for them and maintain good communication. However, in order to understand this knowledge, scientists first have to acquire it and analyze it (Lamers and Fiel, 1995).

In order to characterize a production system, farmers' perceptions of the different technological options are very important, especially the characteristics they use to judge whether a technology is appropriate for them. This judgment does not necessarily consist of accepting or rejecting them, but generally classifying them as less appropriate (Bell6n, 2002 and Lambert, 2010).

The variability identified by farmers on their land should therefore be explicitly taken into account, as it can be a key factor in the development and/or adoption of agricultural technologies. These taxonomies can help identify where technologies are appropriate or not. In addition, knowing them can be useful for scientists because it helps them to communicate better with farmers (Bell6n, 2002).

2.2.2 Characterization of Angola's agro-productive systems

According to Sardinha and Carriço (1975), in the period before independence, agriculture was the strongest sector of the economy, serving as the basis for all the country's economic development and providing sustenance for more than 85% of the population. In this period, the area under maize cultivation was 1,459.3 million hectares. It was also the world's fourth largest coffee producer and a major corn exporter. The production of cereals, maize, millet, millet and rice, legumes, beans, peanuts, roots and tubers, cassava, reindeer potatoes, sweet potatoes and yams, and other industrial crops such as cotton, sisal, sugar cane and oil palm also contributed to domestic and foreign trade (MINADER, 2009).

Agriculture in Angola only accounts for 12% of the Gross Domestic Product (GDP), thus increasing food self-sufficiency (MINADER, 2006 and FAO, 2011). However, it is a fundamental activity for the Angolan population since, as we mentioned earlier, around 65% of the population depends on this economic activity to survive. The agriculture practiced is essentially of the family type, where the purpose of food production is self-consumption, with surpluses being sold essentially in local markets.

According to the Angolan Ministry of Agriculture and Rural Development (MINADER), among the main crops in terms of quantities and areas cultivated for the 2019-2020 campaign were: cassava, corn, sweet potatoes, potatoes, etc (Table 1).

Table 1. Production, area and productivity of the main crops in the 2019/2020 campaign.

N o.	Culture	Production (t)	Area (ha)	Yield ($t.ha^{-1}$)
1	Corn	743 000	1 094 000	679
2	Massango/ massambala	138 000	353 000	391
3	Rice	8 700	6 000	690
4	Cassava	8 607 000	749 000	11 491
5	Beans	109 000	353 000	309
6	Peanuts	66 000	182 000	363
7	French fries	309 000	124 000	2 492
8	Sweet potatoes	664 000	145 000	4 579

Source: Gaudioso (2021).

In terms of area, we can see that the most important crops are maize, which, together with massango and massambala, is the staple food in the Central and Southern regions, and cassava, which is the staple food in the North and is increasingly consumed throughout the country.

Although the production figures for cereals (maize, millet and rice) are much higher than in previous years, they still not enough to meet the country's needs (MINADER, 2005). It should also be noted that these overall figures do not mean that the products are well distributed throughout the country, i.e. due to the poor state of the road network and the inefficiency of the marketing systems, probable production surpluses may not have reached the deficit areas.

The cassava crop, due to its low cultivation requirements, the fact that it can remain in the soil for long periods (being harvested only when necessary) and its high average yield per hectare, is today an important factor in food security in Angola. The high productivity recorded in the table above means that today there is considerable excellence in this product in the country (Vasques, 2010).

As far as the production of the main legumes is concerned, for example the case of beans, the country's needs are still far from being met, with only a third of the country's needs being produced. On the other hand, around of peanut needs are met, which is why it is likely that the country will become an exporter of this product in the future. With regard to sweet potato production, it is estimated that there is a surplus of this product in order of one fifth of national production, while reindeer potato production meets around 80% of national needs (MINADER, 2005).

Agriculture in Angola is fundamentally of the family type, where the use of advanced technology is scarce, and manual tools are normally used for land preparation, weeding, sowing and harvesting. The use of agricultural inputs is low, and yields are usually low when compared to world production. It is estimated that traditional farmers (or peasants) are responsible for 80% of national production. The remaining 18% of production comes from medium-sized farmers, and only from commercial farmers (Pedro, 2008).

Having once been one of the world's largest exporters of coffee and other agricultural commodities such as cotton, sisal, maize, manioc and bananas, Angola's agriculture today is characterized by agricultural production far beyond its potential, forcing the country to spend large sums of money on food imports (Agro-negócio, 2014). It should be noted that there are measures and efforts being made to combat poverty and food insecurity in several African countriesincluding Angola. On July 9, 2002, the African Ministers of approved the Detailed Program for the Development of African Agriculture (PDDAA) in NEPAD (New Partnership for Africa's Development).

This program aims to restore agricultural growth, rural development and food security in the African region. The PDDAA comprises five pillars:

1) Expansion of the area under sustainable land management systems and safe control of water resources;
2) Improving rural infrastructure and trade-related capacities for better access to markets;

3) Increasing food supply and reducing hunger;
4) Development of agricultural research, dissemination and adoption of technologies to support long-term productivity growth;
5) Sustainable development of livestock, fisheries and forest resources (Neto, 2008).

It is with this in mind that the Angolan government, with the aim of reducing poverty and food insecurity, asked the FAO to create a National Medium-Term Investment Program (PNIMP) and a portfolio of Bankable Projects (BIIP, s) (MINADER, 2004).

The viability of investments in the sector is affected by various constraints. Dependence on the foreign market and the inflated economy increase the costs of agricultural production and equipment, which is mostly imported, and there is also the aggravating factor low productivity levels in most products (Angola - Profile of the Country's Private Sector, 2012).

The extent of Angola's territory and the existence of favorable natural conditions, particularly soil and climate conditions, represent extraordinary potential for the country's agricultural and forestry activities. Although the national economy is mainly supported by non-renewable natural resources, the agricultural sector makes an important contribution to the country's economy, being the third largest contributor to the country's GDP, as well as the main reservoir of labor par excellence (Ministry of Agriculture, 2013).

According to Agro-negócio (2014), investments in production systems in the agricultural and livestock sectors are medium- to investments. Even when it comes to annual crops, there is a need to optimize production systems that affect productivity, such as variety selection, fertilization adjustment, soil correction, pest and disease control, etc.

In this regard, the Angolan Executive, through the Ministry of Agriculture, has been developing programs aimed at promoting and fostering private economic activity, fundamentally the production of goods and services, directly and indirectly linked to agricultural production (Ministry of Agriculture, 2013).

Climatic characteristics, knowledge and generational experience are key factors in the process of developing productive sectors that promote the dynamization of the agro-industrial and logistical sector, taking advantage of the benefits offered by the province's geostrategic position, offering possibilities for supplying the national market, replacing imports and boosting the development of the meat sector and its derivatives (Ministry of Agriculture, 2013).

2.3 Producer characteristics and socio-land and economic aspects

2.3.1 Schooling and associations

According to Freitas and Bacha (2004), farmers' higher level of education is related to their entrepreneurial skills, making it easier for them to adapt to cyclical and structural changes in agriculture. Given the importance of schooling, the authors positively correlate human capital with the level of growth in the states. Reinforcing these inferences, Wizniewsky and Wizniewsky (2009) state that low levels of schooling have a negative impact on family farm management by making it difficult to implement certain technologies and understand sustainability. However, they emphasize the importance of accumulated knowledge for conducting activities on the farm.

According to Moraes and Curado (2004), the aim of rural community associations is to improve the production process, as well as community, by integrating the actions of family farmers. The practice of solidarity can also facilitate relationships with professionals linked to the environment, institutions and society as a whole (Lazzarotto, 2002).

2.4 Peculiarities of vegetable production

The most striking feature is the intensive use of the soil, cultivation, labor and modern agricultural inputs (seeds, pesticides and chemical fertilizers). These inputs are used in large amounts per cultivated area. On the other hand, it allows for high net incomes per cultivated area (Filgueira, 1981).

The olive grower is the type of rural entrepreneur who makes the highest profits per unit of area farmed compared to other farmers or breeders. This is because, in most cases, the growing cycle for vegetables is much shorter for other crops. As an example: in one year, on the same plot of land, you can use 3 transplanted tomato crops, or 6 transplanted lettuce crops, or 12 straight-planted radish crops. The cycle of vegetables is usually 3 to 6 months, with exception of snapper (which is perennial) or chayote (semi-perennial) (Makishima, 1983).

As the areas are smaller, we can improve the intensive cultivation, using manual pollination, fumigation of the beds, production of seedlings in containers, thinning of the fruit, foliar fertilization, etc. This makes intensive use of labour and land. Due to its high physical and economic profitability, olericulture allows the use of land with low natural fertility, which would be uneconomical for other crops (Zamberlam and Froncheti, 2001).

2.4.1 Classification of vegetables

According to Bevilacqua (2002), due to the large number of species involved and the particularities of each crop, there is a need a methodology capable of highlighting the botanical and technological similarities and differences between these crops.

For this reason, we try to group them together in a didactic way and, in this sense, there are various classifications based on common characteristics. One very old classification considers the parts used in human food and which have commercial value as the criteria for grouping. This classification currently being used, with minor modifications, by the National System of Supply Centers (Zamberlam and Froncheti, 2001):

- ✓ Tuberous vegetables: those whose usable parts grow inside the ground, including: tubers (potatoes, yams), rhizomes (yams), bulbs (onions, garlic) and tuberous roots (carrots, beet, sweet potatoes, cassava).
- ✓ Fruit vegetables: the fruit is used, green or ripe, in whole or in part: watermelon, peppers, okra, peas, tomatoes, jiló, eggplant, pumpkin.
- ✓ Herbaceous vegetables: those whose usable parts are above ground, tender and juicy: leaves (lettuce, taioba, cabbage, spinach), stems (asparagus, fennel, celery), flowers and inflorescences (cauliflower, broccoli, artichokes).

2.4.2 Origin and geographical distribution of cabbage cultivation

Several studies conclude that the cultivated types of Brassica oleracea originated from a single parent similar to the wild form, which was taken from the Atlantic coasts to the Mediterranean, and that the evolution and selection of the different cultivated types took place in the eastern Mediterranean. Brassicas are cosmopolitan species, particularly abundant in the Mediterranean, Southwest Asia, Central Asia and the West Coast of North America. Among the species economic interest is rapeseed (Brassicanapus) and some ornamental genera (Cleome, Hesperis, Erysimum, Iberis, Lunaria, Lobularia) (Almeida, 2014).

At first, cabbage cultivation was concentrated on the Italian Peninsula and, due to intense trade relations in Roman times, it spread to different areas of the Mediterranean. During the 16th century, cultivation spread to France and appeared in England in 1586. In the 17th century it became widespread throughout Europe, and finally during the 19th century the European colonial powers extended cultivation to the whole world (Triani *et al.*, 2015).

After the discovery of the New World, Europeans introduced it to America and it became very popular with the indigenous people. As early as the 17th century, Nicholas Culpeper, a lover of alchemy, physics and astrology, suggested in his work "Complete Herbal" that cabbage juice mixed with wine could improve the pain of a viper bite. The crop has a good tolerance to soils with high salt and gypsum contents, but is intolerant of other environments where it has to compete with other plants in different conditions (Saavedra, 2013).

Leafy cabbage, an annual or biennial vegetable from the Brassicaceae family, also known as common cabbage and butter cabbage, originated on the European continent. Its consumption has been gradually increasing, probably due to new ways of using it in cooking and recent scientific discoveries about its nutritional and medicinal properties (Bianco, 2015).

The cultivars Manteiga de Ribeirão Pires (A), Manteiga de Jundiaí (L); Manteiga São José (N), Pires 2 de Campinas (R), Orelha de Elefante (U), Vale das Garças (V), Comum (X) and IAC - Campinas - Mendonça stand out in terms of the number of leaves per plant, their size, appearance and taste, pleasing the consumer.

The Brassica genus includes many species of great interest due to their economic importance and nutritional value. Consumption of these vegetables is associated with beneficial effects on health, such as protection against cancer and cardiovascular diseases (Oberley and Oberley, 1997; Moore *et al.*, 2007).

Within this genus, Brassica oleracea is especially important as it includes a wide variety of crops such as sorrel, cabbage, cauliflower, broccoli and Brussels sprouts. Cabbage is particularly rich in antioxidant compounds, such as vitamins, phenols, carotenoids and glucosinolates, and is also rich in calcium (Rosa y Heaney, 1996; Nilsson *et al.*, 2006).

The cabbage plant can be propagated by seeds or shoots. Seeds are sown in isopora trays and then transplanted with a root ball, while sprouts are taken from the leaf axils of adult plants and transplanted into beds and/or trays, this method being more commonly used by farmers. Seedlings of hybrid cultivars are only produced by seeds, as the plants do not produce Trani *et al.* (2015). Propagation using shoots is simpler, as the shoots are planted in beds, which root and grow in the same place.

The leaves should be harvested every seven to ten days, choosing well-developed leaves around 20 to 30 cm long, although there are more demanding markets that prefer leaves 25 to 30 cm long. During harvesting, it is recommended to remove old leaves and shoots to encourage the continuous formation of new leaves on the main stem of the plant (Trani *et al.*, 2015).

2.4.3 Taxonomic classification

Cabbage (Brássica olerácea L.) was first described by Linneo in Species plantarum (1753), describing a wild species. According to Almeida (2014), the genetic designation of the species considers two botanical varieties:

- ✓ Brassica oleracea L., variety Acephyala DC, which includes kale, a cultivar of ornamental kale and forage kale.
- ✓ Brassica olerácea L., variety Constata DC, where the truncated cabbage, also known as kale or Portuguese cabbage, belongs.

Table 2. Taxonomic classification of cabbage (Brassica oleracea L.).

Taxonomy	Nomenclature
Kingdom	*Vegetable*
Sub-kingdom	*Tracheobiont*
Division	*Magnoliophyta*
Class	*Magnoliopsida*
Subclass	*Dilleniidae*
Order	*Brassicales*
Family	*Brassicaceae*
Gender	*Brassica*
Species	*Brassicaoleracea l*
Common name	*Kale*

Source: Gaudioso (2021).

Cabbage belongs to the Brassicaceae family, also known as Cruciferae, which includes more than 300 genera and 3000 species, particularly abundant in the Mediterranean,

Southwest Asia, Central Asia and West Coast of North America. The family also includes several horticultural species great economic importance. There are 22 species of Brassicaceae cultivated as vegetables, belonging to 14 genera (López, 2010).

According to Martínez (2010), the Brassica oleracea crop group is divided into seven crop groups:

- ✓ Brassica oleracea Acephala Group - cabbage.
- ✓ Brassica oleracea Alboglabra Group - kai-lan (Chinese broccoli).
- ✓ Brassica oleracea Botrytis Group - cauliflower (and romanesco).
- ✓ Brassica oleracea Capitata Group - cabbage.
- ✓ Brassica oleracea Gemmifera Group - Brussels sprouts.
- ✓ Brassica oleracea Gongylodes Group - kohlrabi.
- ✓ Brassica oleracea Italica Group - broccoli.

Cabbage is an annual or biennial plant that can reach 2 meters in height during the vegetative phase.

2.4.4 Botanical characteristics of cabbage

Root

Cabbage is considered to be a shallow-rooted plant, with a well-defined pivoting root that reaches up to 80 cm deep. It comes from numerous branches and absorbent hairs and can be thick roots that serve as a storage organ, from which a fasciculated root system originates (Reghin *et al.*, 2008).

Stem

During the first vegetative cycle (germination to cabbage formation), cabbage forms a short, herbaceous, erect stem that varies in size and is cylindrical. According to López (2010), it can reach a height of between 60 and 80 cm and becomes woody as the plant gets older. He points out that the lateral buds are located on the stem, in the axils of the leaves, and during the first stages of the vegetative cycle they are at rest. Only the apical bud is active. Vegetative growth. This growth occurs as the leaves develop and are harvested (Reghin *et al.*, 2008).

Inflorescence

Cabbage produces an inflorescence in the form of a terminal bunch, paniculate or not, the length of the bunch varies from 60 to 80 cm. The flowers are yellow, perfect and regular, medium-sized, hermaphrodite and fertile, mostly allogamous, with four sepals and four white or pale yellow petals, smooth or rough, six woolly threads (2 short outer ones and 4 long inner ones) and a pistil with two cavities (López, 2010).

Fruit

The fruit is a silique capsule, which means that it is a wide and narrow pod, usually descending, with a length ranging from 5 to 13 cm, the appendix of the capsule is short (Reghin *et al.*, 2008).

Seeds with no endosperm, usually in one or two rows, with thick cotyledons, 8 to 9 seeds, glossy black, round and without wrinkles, of medium size. The longevity of the seeds in good storage conditions reaches 5 to 10 years (Reghin *et al.*, 2008).

2.4.5 Cabbage's soil and climate requirements

The quality of vegetables is influenced by pre-harvest factors and others such as genetics and environmental components (Barber, Maddux, Kissel, Pierezynski and Bock). Even in field crops, most environmental factors are difficult to shape and have been shown to have a major influence on the quality and nutritional value of many agricultural products (Romojaro *et al.*, 2007).

The climatic factors together are fundamental to the proper functioning of crops, since they all closely related and the action of one of them affects the rest. Cabbages are cool-season crops, moderately tolerant of frost.

Temperature

Almeida (2014) stated that temperature is the main climatic factor for the growth and development of the crop. Basically, the best lateral growth can be stimulated by suppressing the apical bud, making possible to form heads, although they are of little commercial importance.

Leaves

The leaves of this crop are large, usually 45 cm long by 35 cm wide and slightly petiolate, alternate, usually opposite, semi-thick, light green in color, more or less oblong and oval or almost circular in shape, with a wavy edge. The leaf surface is covered in sepicuticular wax which makes it difficult for water to remain on it, causing it to run off. The rosette that forms the leaves varies greatly in diameter, depending on the variety, and can range from 50 cm to 1 m.

The number of leaves varies from 10 to 15 in early varieties, 20 to 25 in intermediate varieties and 25 to 30 in late varieties. Some leafy varieties can reach a height of about 2 meters during the phase and produce more at temperatures ranging from 15 to 18 °C, with the optimum being 18 to 24 °C. Climates that are influenced by large bodies of water, with cool, uniform temperatures, are favorable for cultivation. However, plants can withstand low temperatures, and it has been observed that temperatures of 3 °C severely damage small plants with 2 to 4 leaves, resulting only in a delay in final production, without altering quality. On the other hand, if the low temperatures occur when the plant has 8 to 10 leaves, the cold induces flowering, depending on the variety.

Temperatures above 30 °C are harmful to this vegetable, especially in conditions of low soil and air humidity, affecting it less when there is a relative humidity over 80% (Sousa and Resende, 2010). The optimum temperature for germination is between 18 and 20 °C. During

the growing phase of the crop requires temperatures of between 14 and 18°C during the day and 5 to 8°C at night, as cabbage requires a difference in day and night temperatures (Infoagro, 2008).

Humidity

In growing conditions, temperature, light intensity and relative humidity are often correlated. If one of these factors is lower than the optimum value, it can lead to poor plant growth. Cabbage is a high humidity plant, so ideally the humidity should be above 70%. In drier areas, the soil should not dry out so that a humid microclimate can be established (Ponce, 2010).

The root system of cabbage is very small compared to the aerial part, so it is very sensitive to a lack of humidity and can withstand a bad period of drought, even if it is very brief. At the beginning of the plant's development, the relative humidity should be between 65 and 75 %. Relative humidity above 75% encourages the development of diseases, especially those caused by fungi and bacteria. Thus, certain pests are favored by high temperatures and low relative humidity (Sousa and Resende, 2010).

It has also been observed that relative humidity is one of the factors that affects the quality of the leaves, specifically their texture, along with other factors such as soil moisture, nutrient availability and temperature (Ponce, 2010).

Relative humidity also has an important effect on the accumulation of calcium (Ca) in the leaves due to its translation, which takes place with the flow of mass in transpiration. Factors that reduce evapotranspiration by the plant's organs also reduce calcium accumulation (Wikipedia, 2009).

Calcium is a fundamental element in the plant's mechanical tissues, as it affects cell enlargement and, in the event of a prolonged deficiency, it can reduce growth and even cause leaf necrosis. This element also affects the quantity and organoleptic qualities (Sousa and Resende, 2010).

Luminosity

Brightness is another important factor, especially during the initial growth period. The productivity of the cabbage crop, as well as its color and texture, depends largely on sunlight. This crop prefers low light. If it is to be sown in full sun, it should be done at the beginning of summer and in mid-summer, protected from direct sunlight with a shade net or other device (Mankin and Fynn, 2006).

For this reason, the location of our country is ideal for this type of crop. In some tropical and subtropical regions it grows well, as long as it is in high areas, and it can behave as a perennial due to the absence of winter in these regions (Sousa and Resende, 2010).

The absorption of water and certain nutrients such as nitrogen and potassium increases in response to increased solar radiation. The maximum absorption rates were observed during the brightest part of the day (Mankin and Fynn, 2006).

Soil

Sousa and Resende (2010) state that vegetables are successfully grown in soils with an adequate amount of nutrients and water. These conditions are necessary for the vegetables to remain soft and juicy, which are important requirements for good product quality. The requirement for fertile soil vegetables is because they have a poorly developed root system (reduced absorption area).

Kristaponyte (2015) reports that cabbage prefers sandy loam soils with good moisture retention capacity, fertile, shallow 40 to 60 cm, well-drained and with good organic matter content. The root only takes advantage of the upper part, so deep soil preparation is not beneficial. It is a plant that copes well with calcareous soils and mildly saline water, although it does prefer quality soils and water. Soils that produce waterlogging are unfavorable.

Normal soil pH values for this crop range from 6 to 7, although they are slightly tolerant of acidity, with a pH range of 5.5 to 6.8 so nutritional deficiencies don't occur and diseases don't proliferate, and they are reasonably tolerant of salinity, with an optimum pH of 6.2 to 6.5. They grow well in any type of soil, from sandy to heavy due to their fine radical branching (Faxsa, 2008), among other causes.

2.4.6 Fertilizing cabbage

Barber *et al.* (2012) showed that cabbage is a species that responds satisfactorily to the contributions of well-decomposed manure incorporated into the previous crop. The contribution of mineral fertilizers varies according to the cycle of the varieties to be grown. To cover the nutritional needs, expressed in kilograms per hectare, the following ranges can be considered: 150 to 350 of nitrate, calcium ammonium, 70 to 120 of calcium super-phosphate and 200 to 300 of potassium sulphate. As for other nutrients, deficiencies or excesses of magnesium, potassium or calcium should be rectified.

Caseiro and Filho (2005), reported that cabbage demands a lot of nitrogen, so it should be applied according to the determination of the analyses, the most appropriate being to use stabilized nitrogen, as it reduces the concentration of nitrate in the leaves by approximately 15%.

With phosphorus, you have to be very careful: a higher dose than necessary will precipitate flower growth. The role of potassium is essential for achieving a good quality harvest, as it provides tolerance to hostile environmental conditions such as drought, frost or disease. A lack of potassium causes shortening of the internodes and gives a violet pigmentation to the veins of the leaves (Barber *et al.*, 2012). Boron deficiency should also be monitored.

Kristaponyte (2015) reports that nitrogen is important for growing cabbage, especially in the first 2/3 days of its cultivation. The application of nitrogen in the form of stabilized nitrogen reduces the concentration of nitrate in the leaves by 10 to 20%, so stabilized fertilizers are especially suitable for cultivation.

As for microelement deficiencies, cabbage is particularly susceptible to boron and molybdenum deficiencies.

A fertilization program recommended by Almeida (2014) for growing cabbage would be:

Bottom dressing:

- ✓ 14 to 24 $t.ha^{-1}$ of manure or fermented chicken.
- ✓ 600 $kg.ha^{-1}$ of NPK complex (15 - 15 - 15).
- ✓ 240 $kg.ha^{-1}$ of magnesium sulphate.

Top dressing:

- ✓ 240 $kg.ha^{-1}$ of ammonium nitro sulphate 10 to 20 days after planting.
- ✓ 300 $kg.ha^{-1}$ of potassium nitrate 30 to 40 days after planting.
- ✓ 240 $kg.ha^{-1}$ of ammonium nitro sulphate when the vegetation completely covers the soil.

Pereira *et al.* (2013) pointed out that it is necessary to use 5 $kg.m^{(-)(2)}$, mix well and leave to stand for a week, then 30 hours before sowing add complete fertilizer.

As for organic fertilization, many vegetables, such as lettuce, cabbage and kale, do not tolerate fresh manure, so they require decomposed or completely decomposed medium (Flebes, 2009).

2.4.7 Pests and diseases affecting cabbage

According to Triani, Passos, Teodoro, Santos and Frare (2015), the main pests and diseases that affect cabbage cultivation are:

- ✓ Defoliator, mockingbird and sweeper worms, which eat cabbage leaves and stems as larvae.
- ✓ The aphid, which is found on the stems and on the reverse side of the leaves, acts by sucking the sap and injecting toxins, turning the leaves yellow and weak, eventually causing death.
- ✓ Leaf miner, which affects the areas near the central nerve of the youngest leaves.
- ✓ Snails and slugs, which eat the leaves and tear them.

Diseases

- ✓ Algae disease, which causes wilting of the plants caused by Rhizoctonia solani and strangulation of the plant's neck.
- ✓ Mildiu, the causal agent of this is Peronospora parasitica, its symptoms are the manifestation of a white fuzz on the reverse of the leaves and on the bundle, chlorosis or yellowing, later the spots on the bundle become dark in color.
- ✓ Cenicilla, caused by the mushroom Erysiphe polygoni, presents a whitish cenicillae on the bundle and the reverse of the leaves.
- ✓ Botritis (Botrytis cinerea) is the cause of tissue rot. Attacks can be found both on the leaves and on the neck of the plants, always with characteristic grey mycelium.

2.5 Nutritional properties and health of cabbage

In cabbage (Brassica oleracea L.), water can exceed 90% of the fresh weight, so its calorie content is very low due to its fat content, but stands out for its protein content (3 to 5%) compared to other vegetables. Kale, like so many vegetables, is very rich in fiber and minerals, as well as calcium and potassium. Its vitamins include A, B1, B2, C, K and folic acid (Infoagro, 2008). It also has an anti-cancer effect due to its glucosinolates (Cartea *et al.*, 2008; Trani *et al.*, 2015).

Flebes (2009) described that the mineral composition of cabbage leaves shows 19.7 mg g^{-1} of calcium (Ca), which is the most concentrated macronutrient, followed by potassium (K) and phosphorus (P), with 13.5 mg $g^{(-1)}$) and 5.73 mg $g^{(-1)}$respectively. Among the micronutrients, the most abundant are iron (Fe), with 72.6 µg g^{-1}, manganese (Mn) with 53.5 µg g^{-1} and zinc (Zn) with 39.4 µg g^{-1}. Other nutrients with a high concentration in cabbage leaves are strontium (Sr), with 252.0 µg $g^{(-1)}$) and aluminum (Al) with 29.3 µg g^{-}

[1]. It should be noted that strontium is absorbed by the human body in the same way as calcium.

Cabbage is a popular consumer product that shares general characteristics with other agricultural and horticultural products. One common feature is its high content of glucosinolates (GLS), compounds of great importance in the flavor and aroma of these vegetables (Almeida, 2014).

In cabbage plants, the main GLSs present are sinigrin, glucoiberin and glucobrassicin, which can vary depending on the development of the plant, with the greatest occurrence being seen in shoots. In addition, it has other properties such as its herbicidal action, due to its inhibitory effect on agermination, as an antifungal and fulfills a defense function against insects and pathogens since it seems to act as a repellent, and attracts certain insects that have a beneficial effect, such as Diaeretiella rapae (Cartea *et al.*, 2008; Kushad, 1999; Moreno *et al.*, 2006; Verkerk *et al.*, 2009; Vilar *et al.*, 2008).

Cartea *et al.* (2008) pointed out that cabbage is characterized by its high nutritional value and antioxidant activity. Its leaves should be eaten raw or after blanching, as cooking the leaves alters the nutritional value and reduces the antioxidant activity of its compounds, especially vitamin C, polyphenols and β-carotene.

Its leaves were probably used by the Egyptians as medicinal plants, while the Greeks served it as public food. During the centuries of Roman domination of the Mare Nostrum, it is mentioned in different works such as "De ré rústica de Cartón", where it appears as a medicine to combat intestinal and lung diseases. The gastronome Apicius would also name it in some of his recipes in "De récoquinaria", in the 1st century AD (Saavedra, 2013).

On the other hand, some of these plants were used in the past as medicinal plants to treat headaches, colds, flu, gout, diarrhea, cuts, wounds and peptic ulcers. They were considered digestive and intoxication-eliminating plants and are used today for the same purpose. Other varieties are very popular as ornamental plants; there are specimens in different colors (white, lilac, etc.) that are very useful for decorating urban gardens, especially during the coldest times of the year (Anonymous, 2015).

(Figure 2).

Knowledge of the nutritional properties of the cabbage plant (Brássicaoleráceae L. var. acephala DC.) is important for obtaining dietary information. Studies carried out by Ayaz (2006) on cabbage leaves showed that the main soluble sugars are fructose (2,011 mg 100 g^{-1}), glucose (1,059 mg 100 $g^{(-1)}$)) and sucrose (894 mg 100 $g^{(-1)}$)), while organic acids are more abundant in citric (2,213 mg 100 $g^{(-1}$)) and malic (151 mg 100 $g^{(-1)}$) acids. Among the fatty acids, alpha-linoleic (54.0 %), linoleic (11.8 %) and palmitic (11.8) acids stand out.

Amino acids include glutamate (12.) and aspartic acid (10.), and the most important proteins are: threonine; valine; isoleucine; leucine; tryptophan; lysine; methionine + cysteine; phenylalanine + tyrosine, the composition of which in cabbage leaves conforms to the standard recommended by the World Health Organization (WHO) (Souza, 2018).

Some varieties are used as condiments, for industrial purposes, to obtain olive oil and in fodder, for animal feed, due to their richness in regulatory substances, proteins and high digestibility, making them an ideal complement for livestock, as well as fresh fodder or in the form of protein concentrates. Its consumption results in less methane gas being generated in the digestive tract of ruminants, which is of great interest from environmental point of view.

Also, like other leafy greens, cabbage varieties a suitable food for diabetics, because their hydrocarbon content makes it easier to use the auxiliary substances, which act in a similar way to insulin. Its cardiovascular benefits are remarkable: it contains "sulforaphane isothiocyanate", a natural anti-inflammatory that acts on the heart, preventing cardiovascular damage, and helps regulate inflammatory processes. Consuming it prevents ischemia, arteriosclerosis and the possibility of suffering a heart attack (Kimoto, 2006).

Due to its high fiber content and as a source of amino acids, one of its most notable benefits is that it is considered an anti-cancer food, due to its high content of glucosinolates, substances that require enzymes to fight free radicals, which ultimately lead to the appearance of cancers. But it is only recently that studies carried out in Japan and USA have shown that cabbage is really effective in preventing certain types of cancer, such as colon cancer and hormone-dependent cancers such as breast and ovarian cancer, as it stimulates women's metabolism (Anonymous, 2015).

Some cabbage varieties have smooth leaves, while others have striped or crinkled leaves (Cartea, Velasco, Obrégon, Padilla and De Haro, 2008). Consumers usually prefer the striped-leaf varieties to the varieties. However, the smooth-leaved variety is more common, as it is grown as a fodder crop (Vilar *et al.*, 2008). Some of the nutrients in Brassica oleracea var. acephala cv. Galega have been quantified in previous studies (Martin, 2006).

The leaves are sold in bundles in units, kilograms (kg) and minimally processed in packaged portions. Carnelossi *et al.* (2005) found in minimally processed cabbage that slicing increases the leaf's respiration rate twice, and that rapid cooling reduces its metabolism, but without increasing storage time. The authors also concluded that harvesting time influences the respiration rate and the activity of polyphenol oxidase 3 (PPO) enzymes, with harvesting at 7am being recommended.

2.5.1 Nutritional composition of cabbage

The composition of cabbage (Brassica oleracea L.) can vary due to both extrinsic and intrinsic factors. Among the intrinsic factors, the most important is the genotype, which is responsible for differentiating species, subspecies, varieties, etc. The age of the plant is also very important. Extrinsic factors are diverse and include growing conditions (climate, soil properties, use of irrigation water, use of fertilizers, application of phytosanitary products, etc.) and post-harvest treatments (form and timing of harvest, application of conservation treatments, etc.) (Almeida, 2014). Elements found in cabbage leaves (Table 3).

Table 3. Nutritional composition of cabbage (Brássica olerácea L.).

NUTRIENTS	ONE	QUANTITIES
Energy	Kcal	27 - 31
Water	%	91 - 92,4
Carbohydrates	%	9,9
Protein	%	1,4 - 2,0
Fat	%	0,10
Fibers	G	2,1 - 3,1
Iron	mg	35 - 47
Sodium	mg	18 - 28
	mg	23 - 42
Magnesium	mg	15 - 28
Potassium	mg	230 - 246
Ascorbic acid	mg	31 - 57
Vitamin A	UI	126 - 1116
Vitamin B_6	mg	0,19
Vitamin C	mg	35

Source: Gaudioso (2021).

- ✓ Amino acids: involved in the production of antibodies.
- ✓ Arginine: essential for ammonia elimination, tissue repair and muscle building.
- ✓ Ascorbic acid: essential for preventing diseases such as scurvy.

- ✓ Cystine: helps hormonal function.
- ✓ Glutamic acid: improves mental conditions.
- ✓ Niacin: helps with hypertension and lowers cholesterol.

2.5.2 Growing system

According to the Bahia Micro and Small Business Support Service (Sebrae, 2018), among the main factors influencing horticultural activities are: the climate, seasonality of production, non-uniform products, high production costs for some products, risks of losses due to pest attacks, diseases and product price variations. The sector has its own characteristics and depends on factors that often cannot be controlled by the horticulturist.

The same author explained that cabbage can be grown using various systems, the choice of which must depend on the need and availability of inputs, the size of the area and the labor available on each property. Among the most commonly used growing systems are conventional (field), organic, hydroponic, slabs5, Mu Ching and others. The main forms of cultivation are presented below.

III. MATERIALS AND METHODS

3.1 Location and stages of the experiment

The experiment was conducted at the R. L Cantinho do Chicote Cooperative (N.J.F), in Mukolongodjo commune, Cuvelai municipality, due to the good conditions of the chosen area. This municipality is located approximately 34 km from Cunene province. It is 19,270 km long^2, bordered to the north by the municipalities of Ombreira and Cuvang, to the east by the municipalities of Cuchi and Menongue, to the south by the municipality of Cuanhama and to the west by the municipalities of Omnadja and Matala. The experimental stage ran from February to June 2023.

The province is located in the far south of Angola, between latitudes 15° 10' 00" to the north, 17° 24' 00" to the south and longitudes 13° 02' 00" to the west and 17° 23' 00" to the east, with irregular rainfall not exceeding 750 mm per year (Ministério de Administração do Território, 2019).

3.2 Experimental design

The experiment was carried out in open field conditions using a randomized block design, with two treatments each with three replicates (Table 4), for a total of 6 experimental plots in the open field. The total area was approximately 88.0 m^2 (0.008 ha), which was measured using an industrial tape measure and wooden stakes as a guide. Each bed was 1.13 m wide and 5.65 m long, with 0.45 m between the beds.

Table. 4 Treatments used in the research.

Treatments	
T1	Cabbage (Brássica Olerácea L. *var. troncuda or Portuguese*)
T2	Cabbage (Brassica Olerácea L. *var. butter* I -1811 *(B)*)

Twelve plants per treatment were randomly selected as a sample. For this purpose, the plants in the central rows were chosen, leaving the outer rows and one plant at either end of each row as an edge effect, as recommended by Morales and Peralta (2009).

3.3 Cultivar used in the research

Leaf cabbage (*Brassica oleracea*) was used for the experiment: variety Manteiga I-1811 (B) and variety tronchuda o português, given its productive potential, adaptability to the soil and climatic conditions of the area and the demand of the population. The seed was purchased from Shorprite in Namibia in an aluminum-coated jar. No pre-germination treatment was used for sowing.

3.4 Organic fertilizers used

Cattle manure was used at a dose of 30 $t.ha^{-1}$ at the time of transplanting in each treatment. The manure used (cattle manure) was compiled in the cooperative's own barn and, after being harvested, was placed in the shade and moistened to encourage further decomposition. When they were collected, they had to be as homogeneous as possible, have a pleasant aroma and be black in color.

3.5 Physico-chemical characteristics of the soil

In order to know the soil fertility parameters in the study area, reference was made to the analyses carried out by Ricardo (2018) and cited by Victoria (2023), for an area of sandy soil type where this research was carried out (see table 5 a and b). This was an important aspect in determining the doses used in this study.

Table 5. a. Soil chemical attributes.

K^{+} (mg/kg)	Ca^{2+} (mg/kg)	$Mg^{(2+}$ (mg/kg)	And so. (mg/kg)	pH (H_2O)
20	195	19	3	7,3

Table 5. b. Soil chemical attributes (Continued).

Organic Matter (%)	In^{+} (mg/kg)	Fe (mg/kg)	Total Nitrogen (%)
0,2	12	13	0,03

3.6 Ground preparation

The land was prepared in the traditional way using a pickaxe and hoe to make it easier to remove weeds. A lady who lived near the study area was hired to do this.

A first organic fertilization was carried out one month before sowing and at 10 and 35 days after transplanting, with cattle manure at a dose 5 kg/m^2, and goat manure was also applied at the same doses.

3.7 Preparing the nursery

Once the land had been cleared, the nursery was prepared near the experimental area where the beds were made and, with an average level of humidity, the seeds were deposited in a hole made with the finger and once they had germinated, 25 days later the seedlings were transplanted.

The seedlings were transplanted when they were an average of ± 8 to 9 cm high. To facilitate this operation, the nursery was previously moistened and then the poorly shaped plants were eliminated, then the seedlings were carefully removed.

3.8 Growing care during research

Watering was carried out taking into account different times of the growing cycle and depending on the intensities and technical standards of sun cultivation, it had to be watered every two days.

The plantation was kept clean throughout the cycle by hand. Pest control was carried out during all stages of cultivation. A barrier of maize (*Zea mays* L) was used; leaf watering with a liquid solution made from repellent plants such as Neem (*Azadiracta indica*) at a ration of 1L/backpack was also applied 40 days after sowing.

3.8.1 Retransplantation

Re-transplanting took place 5 to 6 days after transplanting, i.e. placing seedlings in place of those that hadn't rooted.

3.8.2 Deshierbo

In order to keep the area clean and avoid weed infestation, weeding was carried out several times during the crop cycle in the morning before watering.

3.8.3 Harvesting

Harvesting took place when the plants reached commercial maturity, which was 85 days after the first seedlings were sown.

3.9 Variables that were evaluated

The selected plants were measured at 30, 50 and 70 days after transplanting, taking a total of 10 plants from each treatment.

A. Growth variables:

- Plant height (cm): measured with a metric ruler, from the base of the pseudo stem to the tip of the last leaf.
- Number of leaves (U): this was done by visual counting for each treatment.

- Pseudo-stem diameter (cm): measured with a kingfisher.
- Leaf width and length (cm): measured with a metric ruler.

B. Performance variables and their components:

- Fresh leaf mass (g): the leaves were weighed for each harvest using an analytical balance, taking 2 leaves per plant from 10 plants per treatment.

3.10 Statistical analysis of the results

Based on the data obtained, a simple analysis of variance was carried out for significant differences between the means of the treatments, using the Infostad 2012.Version 1.1 statistical package to process the data.

3.11 Economic valuation

The analysis of the technical economic valuation was carried out on the basis of the production obtained in $t.ha^{-1}$, considering the indicators:

- Cost of production: the costs of all plant production activities were taken, determining expenses, irrigation, seeds, organic matter, soil preparation, among others.
- Production value: if I determine , I consider the production of cabbage and its value on the local market.

Selling price = 1 packet of cabbage leaves costs between 100 and 200 Akz on the local market.

- Gain (Akz): determined by the following expression (Carrasco, 1992).

Gain = Production value - Production cost

IV. RESULTS AND DISCUSSION

4.1- Behavior of plant growth variables

4.1.1 Plant height .

From the results obtained in the experiments, it can be seen that there was a statistical difference between the different treatments used to grow the leaf cabbage and the plant height variables (figure 1). This indicates that the best results were obtained in treatment T1 (cabbage (Brássica Oleràcea L. *var. troncuda o português*)).

Figure 1. Plant height variables.

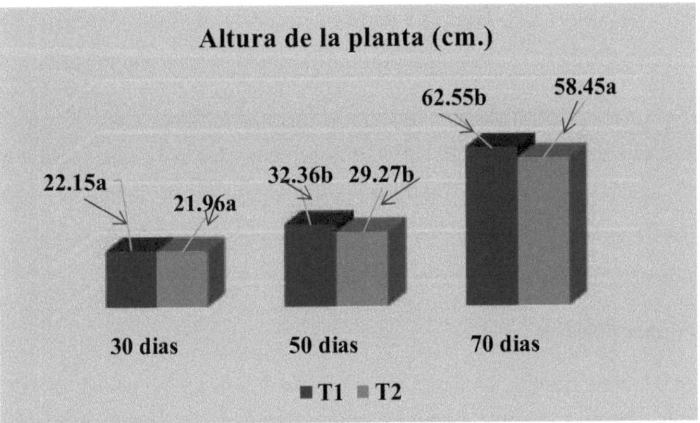

Equal letters in the same column do not differ statistically according to Duncan's test for $p \leq 0.05$.

At 30 days of age, there was no significant difference between the treatments, where the plants reached 22.15 and 21.96 cm in height. At 50 days, the height of the plant showed the most in treatment T1 (32.36 cm), and at 70 days it reached 62.55 cm.

It should also be noted that from the fifth day after transplanting, there was a change in the color of the leaves, which shows that the plant absorbed the nutrient quickly. In this sense, Silva (1985) and Lyons *et al.* (1994) observed significant differences when high levels of nitrate were applied to plants produced with manure and mineral fertilizers. This seems to be partly due to nitrogen becoming available to plants more gradually (Rodrigues, 1990 and Kiehl, 1993).

The results were compared to those found by Steiner *et al.* (2009), who evaluated the production of butter cabbage with organic compost, and it was noted that the values at height were higher than the results found by the author, which suggests that cattle manure applied individually provides good plant height development for organic production.

Furthermore, it is important to clarify that in both cases organic cattle manure is used, as I explained earlier, and humic substances are usually applied to the soil and positively affect its structure and microbial population, as well as increasing the solubility of nutrients in the soil. They thus promote greater plant growth, caused by the presence of substances with functions similar to plant growth regulators, and reduce the effect of water stress on plants (Sediyama *et al.* 2000).

According to Silva Filho *et al* (2002), the main functions of these humic substances are to increase the soil's CEC, aggregate particles, reduce bulk density, increase the soil's moisture retention capacity, complexation and chelation, mining and the soil's biological structure

In the past, farmers preferred taller butter cabbage plants due to the ease of harvesting the leaves (Nieuwhof and Pantano, 1969). However, nowadays, the preference for smaller genotypes is growing (Azevedo *et al.* 2012), showing that the values found in this production with organic amendments can be satisfactory for the producer.

4.1.2 Number of leaves

With regard to the results for the number of total leaves (NFT) shown in figure 2, a significant difference was found between the treatments. It should be noted that at 30, 50 and 70 d.a.s., the best results were obtained by the treatment (T1 Cabbage (Brassica Oleracea L. *var. troncuda o Portuguese*) in all cases. These results are due to the better physical characteristics in terms of apparent and particle densities presented by the amendments, which provide greater leaf formation in the plant.

Figure 2. Number of leaves.

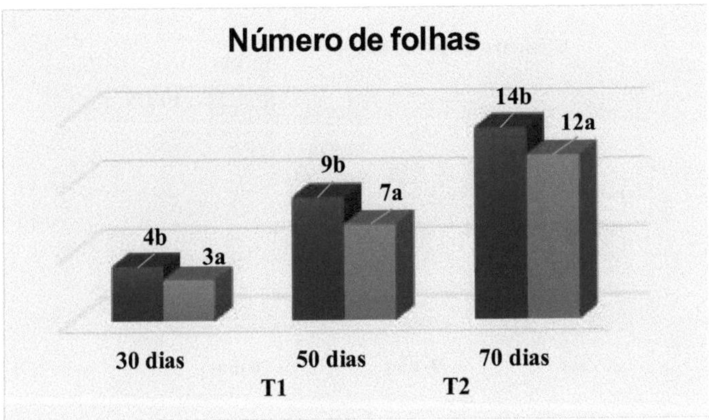

Equal letters in the same column do not differ statistically according to Duncan's test for $p \leq 0.05$.

The total number of leaves represents the vigor of the plant, i.e. it is the maximum expression of the genetic potential added to the environmental growing conditions, and in this case it is desirable that the greater the total number of leaves, the greater the number of commercial leaves. In this study, the best vigor was obtained in treatment T1, which had the best nutritional characteristics, corroborating Milner (2006) who describes that plant growth is greatly influenced by the physical and chemical characteristics of fertilizers.

The results are comparable to those of Taiz *et al.* (2017), who described that leaf senescence is a genetically regulated process that cannot be completely annulled or increased by any mutation, treatment or environmental condition, in which factors such as developmental age or chronological age can be determinants of leaf senescence.

Zeiger *et al.* (2017) in studies carried out with ten cabbage genotypes, showed that water deficit negatively influenced their biomass, production and nutritional quality, with a reduction of 19 to 35 % in biomass depending on the genotype, and loss of nutrients in cabbage plants, including K, P, Fe, Mn and Cu.

4.1.3 Pseudostem diameter (mm)

There was a significant difference ($p<0.01$) for the treatments, time and interaction between them for pseudostem diameter (figure 3).

Figure 3: Response of pseudostem diameter to different treatments.

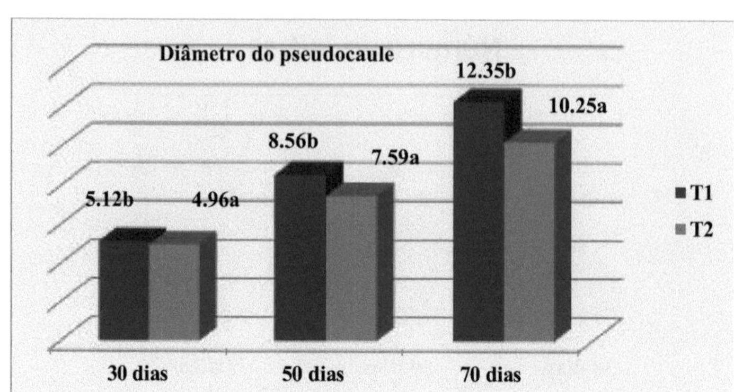

Equal letters in the same column do not differ statistically according to Duncan's test for $p \leq 0.05$.

The average growth in the pseudostem diameter of cabbage plants was linear over the cultivation period (30 to 70 days) for all the treatments evaluated. The largest pseudostem diameter was 12.35 mm at 70 dds in treatment T1 (Cabbage (Brássica Olerácea L. *var. troncuda o português*). The smallest plant diameter at 70 dds was obtained in treatment T2 Couve (Brássica Olerácea L. *var. manteiga* I -1811 *(B)*) 10.25 mm.

Related to this, Taiz and Zeiger (2013) pointed out that stem growth, although less studied than leaf expansion, is also affected by the same conditions that limit leaf growth during periods of low water availability.

Novo *et al.* (2010) obtained linear growth in the stem diameter of cabbage plants up to 80 days for the cultivars Verde-Escura, Orelha de Elefante and Vale das Garças, but after this period, there was a reduction. This reduction was not observed in this study.

Knowing the diameter of the stem is important for defining the ideal time to support the plant and, as a result, reduce plant losses due to wind-driven toppling (Novo *et al.* 2010). However, Trani *et al.* (2015) pointed out that in addition to climatic conditions, the physiological differences in cabbage may be related to the botanical characteristics of the plant and its responses to cultural treatments (fertilization and irrigation).

4.1.4 Leaf width and length .

According to the results obtained for leaf width, the statistical analysis revealed a significant difference between the treatments, with the best result coming from treatment

T1 (Cabbage (Brassica Olerácea L. *var. troncuda o português*)), with leaves measuring 18.9 cm (figure 4). The lowest result was obtained in treatment T2 Cabbage (Brassica Olerácea L. *var. manteiga* I -1811 *(B)*), with results of 16.9 cm. The same happened with the behavior of the leaf length variable, where treatment T1 had a significantly higher result according to the statistical data with values of 28.3 cm and treatment T2 with values of 22.0 cm in leaf length respectively, as shown in figure 4.

Figure 4 Leaf width and length

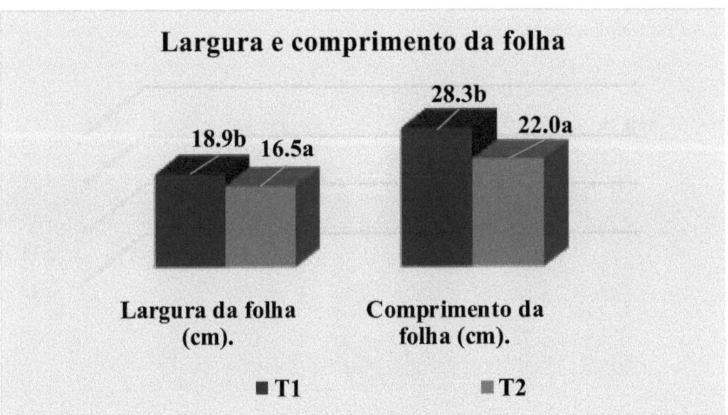

Equal letters in the same column do not differ statistically according to Duncan's test for $p \leq 0.05$.

This information is similar to that described by Taiz *et al.* (2017), who reported that during leaf senescence nutrients are remobilized via the phloem from the source leaves to the drains of vegetative or reproductive growth, and that nutrients from the decomposition of carbon and the conversion of chlorophyll, proteins and other macromolecules can be translocated to other parts of the plant, such as organs in vegetative growth, seeds or developing fruit.

It is important to note that according to Castro *et al.* (2005) old leaves serve as a source of water for new leaves, as they have a high soaking pressure and high water absorption due to their poorly vacuolized cells. Therefore, the senescent leaves of cabbage plants promoted the remobilization of nutrients and water for the new developing leaves.

The positive results obtained in the cattle manure and chemical fertilization treatments are in line with Kampf and Fermino (2000) who published that density is an important characteristic in the composition of the substrate, as it interferes with the porosity and water availability of the substrate, limiting plant growth.

4.2 Leaf weight response of the plants.

Figure 5 shows the plant's response to the average leaf weight per treatment after transplanting, where the highest results were achieved in treatment T1 (*Portuguese* cabbage (Brássica Olerácea L. *var. troncuda*) with a weight of 420.7 g; 422.9 g; and 423.8 g at 30, 50 and 70 days respectively

Figure 5 - Plant leaf weight response

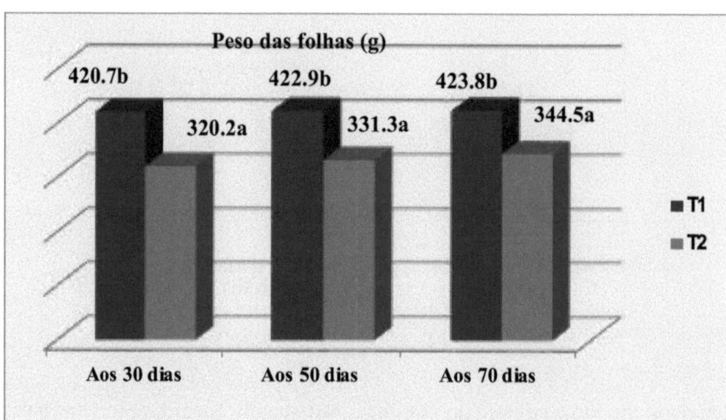

Equal letters in the same column do not differ according to the multiple mean comparison test. Significant at probability ($p<0.01$).

The average weight of the leaves is a component of the yield and is a very important indicator of the quality of the cabbage leaf. They are well related to each other, as they give the leaves a good presence (Huerres and Caraballo, 1996). The weight of the leaves also determines the yield of this crop, regardless of the conditions evaluated.

The result is also reflected in the soil without organic fertilizer treatment, which had relatively little humic substance content, as well as in the formation of stable macro-aggregates in the soil water, among which are pores that ensure aeration and drainage.

The results presented here show that organic matter improves soil fertility, regulates the chemical, biological and physical processes taking place in the soil and affects yield components. In terms of chemical processes, organic matter helps to supply plants with essential elements, which can lead to better leaf width and length (Sousa and Resende, 2010).

Work carried out by various authors has managed to increase the average weight of leaves per plot in leafy vegetables by applying organic matter. This coincides with what Ernesto (2018) says when he states that organic fertilizers are safe, economical and effective systems for obtaining short-term results.

4.3 Economic valuation of research

Economic evaluation uses quantitative methods to estimate the costs and benefits of agricultural research activities (see table 6). It can be done in terms of market prices, costs and the real benefits that producers receive (Falcony, 1994 cited by Dongua 2020).

In order to analyze the economic effectiveness of sowing two varieties of cabbage in the climatic conditions of Mukolongodjo, all the costs incurred in each of the following cultural activities were calculated: transport, irrigation, soil preparation, sowing, cultural care, harvesting and others. To analyze the indicators, the current price of the agricultural product (cabbage) on the shorprite and the real cost of research were used.

Table 6. Economic valuation.

Treatments	**Yield ($t.ha^{-1}$)**	**Sales price (AKz)**	**Production value (Akz)**	**Production costs (Akz)**	**Gain (Akz)**	**Cost of production (Akz)**
T1	3,70	500 000.00	1 850 000.00	61 240.00	1 788 760.00	0.03
T2	2.45	500 000.00	1 225 000.00	59 124.00	1 165 876.00	0.05

Treatment T1 (Cabbage (Brassica Olerácea L. *var. troncuda or Portuguese*)) obtained the best gain with 1,788,760.00 Akz and production cost (0.03 AKz). On the other hand, although treatment T2 (Cabbage (Brassica Olerácea L. *var. manteiga* I -1811 *(B))* had lower production costs (59,124.00 AKz), it obtained a gain of 1,165,*876*.00 Akz and a higher production cost (0.05 AKz*)*.00 AKz), a gain of 1,165,876.00 Akz was obtained a higher production cost (0.05) than Treatment T1 (Cabbage (Brassica Olerácea L. *var. troncuda o português*)), from an economic point of view it is feasible to sow both varieties under the soil and climate conditions in Mukolongodjo.

V. CONCLUSIONS

- ✓ The best results for the variables plant height, number of leaves, leaf width and length, pseudostem diameter and leaf weight were obtained with treatment T1 (Cabbage (Brassica Olerácea L. *var. troncuda o português*)) for the soil and climate conditions in Mukolongodjo.
- ✓ The economic results were positive in both treatments

VI. RECOMMENDATIONS

- ✓ Increasing the area under cabbage cultivation (Brássica Olerácea L. *var. troncuda o português*) in accordance with the soil and climate conditions of the R. L Cantinho do Chicote cooperative in order to increase its economy.
- ✓ Evaluate this crop in other soil and climatic conditions in the province in order to achieve greater production results and increase demand among the population

VII. BIBLIOGRAPHICAL REFERENCES

1. Adib, A. R. and Miranda, C. L. (2007). *Aspects of family farming in Brazil: a literature review.* In: BARRIL, G. A.; CHAVEZ, F. A. (Ed.). Family Farming in the Southern Cone Countries. Assuncion: IICA, 2007. p. 35-81.
2. Agribusiness (2014). *Cooperation network for the agribusiness technologies and services sector.* Market Research & intelligence-Angola.
3. Aguilera M, Bruna G, Brzovic F, Cerda R, Clark M, Chandía A, Domínguez J I, Espinoza A, Faúndez M, García P, Jara C, De Kartzow A, Kern W, Lerdón J, Marchant R, Mora M, Olavarría J, Paillacar R, Quijada A, Troncoso J L y Vargas G. (2003). *Fundamentos de gestión para productores agropecuarios: tópicos y estudios de casos consensuados por universidades chilenas.* Program. Santiago de Chile.
4. Almeida. D. (2nd ed.). (2014). *Manual de culturas hortícola.* Lisbon. Portugal: Editorial Presença.
5. Altieri, M. and Nicholls, C. (2013). *Agro ecología: Única esperanza para la soberanía alimentaria y la resiliencia socio ecológica*, 7(2), 65-83.
6. Amisse, J. (1997). *A importância da Extensão Rural*, In: A Extensão Agrária em Moçambique (1987 - 1997). *Maputo: National Directorate of Rural Extension.*
7. ANGOP (2019). *Angola Press News Agency.* Cereal production reaches 2.9 million tons.
8. Anonymous (2015). *Propiedades nutricionales y terapéuticas de la col (Brassica oleracea L.).* i e h www.ceasacampinas.com.br. Consultation date: February 8, 2021.
9. Apollin, F. and Eberhart, C. (1999). *Analysis and diagnosis of production systems in rural areas.* Methodological guide. Ecuador: Quito, 237.
10. Astier, M., Masera, O. and Galván, Y. (2008). *Sustainability Assessment. A dynamic and multidimensional approach.* Retrieved from http://docplayer.es/14885206-Evaluacion-de-sustentabilidad-un-enfoque-dinamico-y-multidimensional.html.
11. Ayaz, N. D. (2006) *Detection of species in meat and products using enzime-linked inmunosorbent assay.* Journal of muscle foods. Vol 7 (2): p.214-220.

12. Barber, K. L., Maddux, L. D., Kissel, D. E., Pierezynski, G. M & Bock, B. R. (2012). *Corn responses to ammonium and nitrate-nitrogen fertilization*. Soil Sci. Soc. Am. J, 56 (1): pp 166-171.
13. Bellón, M. (2002). *Métodos de investigación participativa para evaluar tecnología*: manual para científicos que trabajan con agricultores (p.106). México, D.F.
14. Bentley, Jeffery W. & G. Rodríguez (2001). "*Honduran Folk Entomology*. Current Anthropology, 42(2), 285-301. Retrieved May 20, 2021, from http://www.jefferybentley.com/Honduran%20Folk%20Entomology.htm.
15. Bettiol, W., Ghini, R., Montes, J., Haddad, A. and Siloto, R. C. (2004). *Organic and conventional tomato cropping systems*. Sci. agric. 61(3): pp 253 - pp 259.
16. Bevilacqua, C. R. Helen Elisa (2002). *Classification of vegetables*. Bianco, M. S. (2015). Agroeconomic viability of cabbage intercropping with 'New Zealand' spinach. Thesis (Doctorate in Agronomy) - Faculty of Agricultural and Veterinary Sciences, Paulista State University, Jaboticabal.
17. Bruner, E. M. (2014). *The agingbAntrhopogist*. Anthropology and Humanism. 39 (1): 27-31. ISSN 1559-9167, only ISSN 1548-1409.
18. Buainain, A. *et al*. (2003). *Family farming and the new rural world*. Sociologias, Porto Alegre, n.10, 2003.
19. Carnelossi, A. C. *et al*. (2005), *Physiological response of minimally processed cabbage leaves*. Hortic. Bras. 23 (2).
20. Cartea, M. E.; Velasco, P.; Obrégon, S.; Padilla, G.; De Haro, A. (2008). *Seasonal variation in glucosinolate content in Brassica oleracea crops grown in northwestern Spain. Phytochemistry*, New York, v. 69, n. 2, p. 403-416, 2008.
21. Carvalho, I.; Ferreira, P.V.; Silva, J. A. B.; Teixeira, J. S.; Oliveira, F. S.; Carvalho, A. P. & Santos, P.R. (2016). *Productive analysis of green corn genotypes intercropped with beans*. Horticultura Brasileira 34: 593-599. DOI - http://dx.doi.org/10.1590/S0102-053620160422.
22. Caseiro, R. F. and Filho, M. (2005). *Evaluation of methods for drying primed onion seeds.Hortic*. Bras. 23 (4): pp 887 - pp 892.
23. Cassinda, M. (2016). *Productive response of bell pepper cultivation (Capsicum annum L.) with different planting patterns in Ondjiva*. (Thesis in option to the degree of Engineer). University of Cuito Cuanavale. Higher Polytechnic Institute of Cunene. Angola.

24. Castilo, B. (2016). *Angola: cereal production registers a deficit of 900,000 tons.* 05 May 2023. Retrieved from: http//www.angop.ao on 15-03- 2023.
25. Castro, I., Díaz, L., Pérez Y., Rodríguez, M., Gómez, L. (2010). *Organic fertilizers: a viable alternative to mitigate or reduce the presence of pest-forming phytoparasites.* Universidad Agraria de la Habana "Fructuoso Rodríguez Pérez and Centro de Sanidad Agropecuaria. La Habana. Cuba.
26. CERF: Central Emergency Response Fund (2012). *Report by the Resident/Humanitarian Coordinator on the use of CERF funds in Angola.* Meihua Clinic (2014). Specialist in Chinese medicine. Located at: http://medicina-tradiconal-chinesa.com/2014/21/nutrición/hortalizas/.
27. Dias, M. A. *et al.* (2010). *Maize seed vigor associated with matocompetition.* Revista Brasileira de Sementes, 32 (2), pp. 93-101.
28. Dufumier, M. (1990). *Importancia de tipología de unidades de producción* agrícola en el análisis de diagnostico de realidades agrarias. En J. Berdegué, M. Dufumier y G. Escobar. Tipificación de Sistemas de Producción Agricola. (pp. 63-81). Chile: Santiago de Chile. Retrieved from: https://doi.org/ISBN. 956.7110-01-07.
29. FAO (2011). *Conservation sustainable use seed systems.* IN: The Second Report on the State of the World's Plant genetic Resources for Food and Agriculture. International Technical Conference on Plant Genetic Resources, Rome, 2011. Available at: <http://www.fao.org/agriculture/crops/core-themes/theme/seeds-pgr/gpa/gpa-update/en>. Accessed May 20, 2023
30. FAO (2012). *Crop prospects and food situation.*N. 2 February 2021. Available at: www.fao.org/docrep/015/a/1990s/a1990s00.pdf.
31. Faxsa, N. (2008). *Col General Information.* Available at: http://www.faxsa.com.mx. Consultation date: January 18, 2021.
32. Febles, J. (2009). *Agroecological strategy for soil conservation.* Paper presented at the Master's Course in Plant Nutrition and Biofertilizers. La Habana. Cuba.
33. Filgueira, F. A. R. (1981). *Manual de Olericultura.* Vol. I. 2nd ed. São Paulo: Agronômica Ceres. 338p.
34. Freitas, C. A. and Bacha, C. J. C. (2004). *Contribution of human capital to the growth of Brazilian agriculture*: period from 1970 to 1996. Revista Brasileira de Economía, 58 (4).

35. Gaudioso N, G. N (2021). *Characterization of cabbage (brássica olerácea L.) production systems in Mukolondjo*. End-of-course work in option for the Degree of Engineer. Option: Agriculture. Department of Human Resources. Mandume ya Ndemufayo University Ondjiva Polytechnic Institute
36. García, E. (2004). *Dinámica e interrelaciones entre las propiedades de un suelo pardo, en dos agroecosistemas*. Thesis in option to the academic title of Master in Agricultural Sciences. Guantánamo, Cuba. pp 32.
37. Garcia, N.Q. (2011). *The land problem*. State and the distribution of land rights: the case of WakuKungo. Contexto. Revista de Estudos Científicos de Desenvolvimento (Luanda), No. 2, pp 01-11.
38. Garcia, N.Q. (2012). *Democratic state of rights*: from Habermas to the "African dilemma". Mulemba- Angolan Journal of Social Sciences (Luanda), vol. 11, No3.
39. Garcia, N.Q. (2015). *Family farming in Angola*: the conceptual pitfalls of dichotomous classification. Mulemba, 5 (10).
40. Provincial Government of Cunene (2018). *Guide for the investigated.*
41. Hernández, A. (2009). *Preserving the soil, an imposing mission*. Retrieved from: http://www.tribuna.co.cu/Etiquetas/2009/abril/21/preservar-suelo.html. Consulted May 16, 2023.
42. Hernández, M.A. and López, B.M.P (2009). *Human development and gender in Mexico 2000-2005: advances and challenges*. En: Indicadores de desarrollo humano y género enMéxico 20002005. United Nations Development Program, Mexico, pp.1-26.
43. Hidalgo, F., Houtart, F. and Lizárraga, P. (eds.) (2014). *Agriculturas campesinas enLatinoamérica: propuestas y desafíos*. Quito, Ecuador, Editorial IAEN.
44. INE (National Statistics Institute) (2015). *Angola 40 years of statistics 1975-2015.*
45. Infoagro (2008). *Growing vegetables*. Available at:http://wwwinfoagro.com/hortaliza/col.htm. Consultation date: April 30, 2023.
46. InfoStat (2002). *InfoStat, version 1.1.* User's Manual. InfoStat Group, FCA, National University of Córdoba. First Edition, Editorial Brujas Argentina.
47. Brazilian Institute of Geography and Statistics - IBGE (2006). *Agricultural census 2006*. Available at: <http://www.ibge.gov.br/cidadesat/topwindow.htm>. Accessed on: May 20, 2023.
48. Kamangira, J.B. (1997). *Assessment of soil fertility status for agroforestry interventions using conventional and participatory methods.* Master's thesis for

degree of Master in Agricultural Sciences, Bunda College of Agriculture, University of Malawi.

49. Kimoto, T. (2006). *Cabbage, cauliflower and broccoli nutrition and fertilization.* Available at: Nutrition and fertilization of vegetables. Consultation date: May 30, 2023.
50. Kristaponyte, Irena (2015), *Effect of fertilization systems on plant nutrient balance and soil agrochemical properties.* Available at: <http://www. researchgate.net/2377/. Accessed May 20, 2023
51. Lambert, G. Tania (2010). *Identificación, manejo y conservación de suelos en la comunidad "La Concepción", a través de métodos participativos.* (Tesis defended in option to the title of Master in Environmental Management). University of Granma. Cuba.
52. Lamers, J. & Feil, P. (1995). *Farmers' knowledge and management of spatial soiland crop growth variability in Niger, West Africa.* Netherlands Journal of Agricultural Science, 43, 375-389.
53. Lazzarotto, J. J. (2002). *Rural associations and their viability*: a comparative case study of two rural producers' associations in the municipality of Pato Branco (PR). In: Encontro da associação nacional dos programas de pós-graduação em administração - ENANPAD, 31. Salvador. Proceedings... Salvador: ENANPAD, 2002.
54. Leyva, A. and Pohlan, J. (2005). *Agroecology in the tropics: examples from Cuba.* Plant biodiversity, how to conserve and multiply it. Aachen: Ediciones shakerverlang, 198.
55. López, G. (2010). *Bioagronomic study of 20 cabbage cultivars (Brassicaoleracea L., Var. Capitata).* Thesis in option to the title of Agricultural Engineer. Guantánamo, Cuba.
56. Lores, A. (2008). *The sustainability of the agroecosystems of the "Zaragoza" community, San José de Las Lajas.* Contribution to the study of agro-diversity. Doctoral thesis for the degree of Doctor of Agricultural Sciences, La Habana, Cuba.
57. Lubamba, M. (2018). *Southern Angola continues to suffer from drought and famine.* Deutsche Welle.
58. Maia, M. R. (2012). *Sustainability and family farming in Vitória da Conquista - BA. 287 p. (Doctoral Thesis), Federal University of Sergipe, São Cristóvão, SE.*

59. *Makishima,* N. (1983). *Small-scale vegetable production.* EMBRAPACNPH, 23p. (Technical instructions, 6).
60. Mankin, K.R. and Fynn, R. P. (2006). *Modelling individual nutrient uptake byplants:* Relating demand to microclimate. Agricultural Systems, 50 (1): pp 101 - pp114.
61. Márquez, G. Mervyn (2012). *Production systems and ergonomics*: a reflection for the debate. Industrial Engineering. Actualidad y Nuevas Tendencias. Vol III, No 9. ISSN: 1856-8327.
62. Martin, C. A., Almeida, V. V. de, Ruiz, M. R., Visentainer, J. E. L., Matshushita, M., Souza, N. E. de Visentainer, J. V. (2006). Omega-3 and omega-6 *polyunsaturated fatty acids:* importance and occurrence in foods. Revista de Nutrição, Campinas, v. 19, n. 6, p. 761-770.
63. Martínez, C.F.P. (2010). *Evaluation of the efficiency of three organic fertilizers with three different dosages in the yield and profitability of lactol cultivation.* Thesis in option to the title of Agricultural Engineer. Riomba, Ecuador.
64. Méndez, V. Lenixia (2014). *Participación social de las mujeres en la toma de decisiones y gestión de microempresas en las zonas rurales de Guerrero*.
65. Research area: *Social entrepreneurship.* Universidad Autónoma de Guerrero, Unidad de Estudios de Posgrado e Investigación. XIX International Congress of Accounting, Administration and Informatics.
66. MINADER (2005). *Report on the 2004-2005 agricultural campaign.*
67. MINADER (2006). *Report on the 2005-2006 agricultural campaign.*
68. MINADER (2009). *Biennial Plan for the Agricultural Sector (2010/2011).*
69. MINADER. (Ministry of Agriculture and Rural Development). (2005). *Evaluation Report on the 2004/05 Agricultural Campaign*. Accessed on February 13, 2021, on the website of: Angola: http://www.minader.org/pdfs/seguranca_alimentar/Relat_Camp04_05%20Revisto.pdf.
70. MINADER-FAO. (2003). *Review of the agricultural sector and the food security strategy for defining investment priorities (TCP/ANG/2907) - Agricultural Production Systems.* Working document No. 07. Preliminary version for comments. Accessed on February 28, 2023, *on the website of: Angolan Ministry Agriculture and Rural*

*Development:*http://www.minader.org/pdfs/fomento/volume_iii/sistema_producao_agricola.pdf.

71. MINADER-FAO. (2004). *Review of the agricultural sector and food security strategy for defining investment priorities* (TCP/ANG/2907) - Development of Irrigated Agriculture. Working document No. 09. Preliminary version for comments. Accessed on May 13, 2023, on the website of: Ministry of Agriculture and Rural Development of Angola:http://www.minader.org/pdfs/fomento/volume_iii/desenvolvimento_agricultura_irrigad a.pdf.
72. MINAGRI - Angola (2019). 2017/18 *Agricultural Campaign Report.*
73. MINAGRI - Angola: Ministry of Agriculture of Angola (2013). *Report of the Multisectoral Technical Mission to the Provinces of Benguela, Cuando Cubango, Cuanza Sul, Cunene, Huíla and Namibe.*
74. Moore, L. E., Brennan, P., Karami, S., Hung, R. J., Hsu, C., Boffetta, P., Toro, J., Zaridze, D., Janout, V., Bencko, V. (2007). *Glutathione S-transferase polymorphisms, cruciferous vegetable intake and cancer risk in the Central and Eastern European Kidney Cancer Study.* Carcinogenesis, 28, 1960-1964.
75. Moraes, Ê. G. and CURADO, F. F. (2004). *The Limits of Associativism in Family Farming in Rural Settlements in Corumbá (MS).* In: Symposiums on the Natural and Socio-Economic Resources of the Pantanal, 4., Proceedings... Corumbá. Available at: <http://www.cpap.embrapa.br/agencia/simpan/sumario/artigos/asperctos/pdf/socio/323SC_CURADO_8_OKVisto.pdf>. Accessed on: April 28, 2023.
76. Muñoz, L. (2018). *Asociación de cultivo en el* Huerto. https://www.agrohuerto.com/asociacion-decultivos-compatibilidad-entre-plantas/ (acceso: 16/05/2023.
77. Muñoz, O., Germani, A., Aguilar, C., Jacinto, A. and Sarmiento, F. (2015).*Descripción de los sistemas intensivos de engorde de corderos en Yucatán*, México.
78. Nova Scientia Neto, J. (2008). *Angola: Agriculture and Food.* Portuguese Institute Development Support, Documentation Center. Lisbon.
79. Neto, Baptista and Cabral (2006). *Angola: Agriculture and Food in Angola.* Agriculture, Natural Resources and Rural Development, Vol. I, Ilídio Moreira (org.), Isa Press, Lisbon.

80. Nilsson, J., Olsson, K., Engqvist, G., Ekvall, J., Olsson, M., Nyman, M. y Åkesson, B. (2006). *Variation in the content of glucosinolates, hydroxycinnamicacids, carotenoids, total antioxidant capacity and low-molecular-weightcarbohydrates in Brassica vegetables*. Journal of the Science of Food and Agriculture, 86, 528-538.
81. North, K. & Hewes, D. (2006). *Seguimiento de fincas para el progreso hacia la sostenibilidad.* LEISA Revista de agroecología, 1 (22).
82. Oberley, T.D. and Oberley, L.W. (1997). *Antioxidant enzyme levels in cancer.* Histology and Histopathology, 12, 525-535.
83. OGE (2011). *Republic of Angola*. Report on the General State Budget.
84. ONU (United Nations Organization). (1995). *Fourth World Conference on Women.* Annex II. Platform for action, Chapter 1, point 1. Beijing, September 1995.
85. Ortega, F. (2009). *Preserving the soil, an imposing mission*. Retrieved May 16, 2023, from http://www.tribuna.co.cu.
86. Pacheco, F.; Carvalho, M. L. S. and Henriques, P. D. (2011). *A contribution to the debate on sustainability in Angolan agriculture.* Proc. 2nd Luso-Angolan Meeting on Economics, Sociology, Environment and Rural Development. Luanda, 311-343. ISBN 978-989-8550-20-0 PDNA Post-Disaster Needs Assessment (20). Drought in Angola 2012 - 2016.
87. Pacheco, Raiza (2018). *La agricultura familiar como un medio para la seguridad alimentaria familiar en la parroquia de Chugchilan, cantón Sigchos, en el periodo de marzo a julio de 2017.* (Dissertation to obtain a degree in Human Nutrition). Pontifical Catholic University of Ecuador. Faculty of Nursing. Human Nutrition Course.
88. Palacios, V. and Barrientos, J. (2014). *Caracterización técnica y económica de los agrosistemas de producción en dos resguardos indígenas del Putumayo (Colombia).*Acta agronómica, 63(2), 91-100. doi: 10.15446/acag. v63n2.29358.
89. Paterniani, E. (2001). *Sustainable agriculture in the tropics.* Estudos Avançados43: 303-326.
90. Pedro, H. A. (2008). *Public Irrigation in Angola: Proposal for a Management Model.* Final Course Work in Agronomic Engineering. Higher Institute of Agronomy. Technical University of Lisbon, Lisbon. 50 pp.

91. Pereira de Oliveira, A., Arnaldo, N., Ovídio, P., Das Suany, M., Antonio, D.G. (2013). *Yield of okra fertilized with cattle manure and biofertilizers*. Ciencias Agrarias. 34 (6): pp 2629 - pp 2636.
92. Pérez B. M. A. (2009). *Rapid agroecological system for evaluating soil quality and crop health.* Methodological Guide. Herramienta para la Gestión de Sistemas Agrícolas desde la perspectiva de la Agroecología.
93. Pérez, A., Saucedo, O., Iglesias, J., Wencomo, Hilda B., Reyes, F., Oquendo, G., Milián, Idolkys. (2010). *Characterization and potential of sorghum grain (Sorghum bicolor L. Moench).* Pastos y Forrajes, vol. 33, núm. 1, enero-marzo, 2010, pp. 1-26.
94. Pierri, F. and Brady (2013). *The Decade of Family Farming at the United Nations.*
95. *Piñeiro, M. (2001). Investment options in the rural economy.* En Echeverria (Ed.). Development of rural economies in Latin America and the Caribbean.
96. Ponce, V. (2010). *Prefeasibility project for the production and export of frozen melon pulp to the Venezuelan market in the period 2011-2020.* Thesis presented at the School of Foreign Trade, Integration and Customs. Venezuela.
97. Pudasaini, S. P. (2011). *The Effects of Education in Agriculture*: Evidence fromNepal. American Journal of Agricultural Economics, Vol. 65, Available at:http://www.perpustakaan.depkeu.go.id/peranian. PDF. Accessed May 2023.
98. Reghin, M., Olinik, J.R., Jacoby, C and De Silva, A. (2008) *Mineral and organic fertilization in cabbage (Brassica oleraea L. Var. Capitata).* Total and commercial production. Horticultura Brasileira. 2 (1): pp 13 - pp 16.
99. Ríos, G. G., Munoz, Clara Isabel, Pérez, C. J. C., Franco, G. and Arango, A. DoraInés, *et al.* (2000). *Characterization of agricultural production systems in the department of Caldas*. Revista Informe Técnico. SDALC.
100. Ríos, H. (2009). *Farmers improve crops.* Fitomejoramiento Participativo. La Habana: INCA. 299 p.
101. Rodríguez, G., Horacio, A. de la L., Lérida, Hechevarría, S., Milanés F. Isabel and Rodríguez, F. C. A. (2008). *Comparative study between monoculture and intercropping in various medicinal plants.* Revista Cubana de Plantas Medicinales, 13(3).
102. Romojaro, F., Martínez, M., Madrid, C., Pretil, M. T. (2007). *Pre-harvest factors determining the quality and post-harvest preservation of agricultural products.*

103. Rosa, E. and Heaney, R. (1996). *Seasonal variation in protein, mineral and glucosinolate composition of Portuguese cabbages and kale.* Animal Feed Science and Technology, 57,111-127.

104. Saavedra, G. (2013). *Introduction to vegetable production. Technical Series: Producción de hortalizas para la República de Guinea Ecuatorial.* Numero Food and agriculture organization of the United Nations (FAO). Subregional Office for Central Africa. Republic of Guinea Ecuatorial.

105. Sambeny, Z., Alves, E., Calegar, G., Tollini, H., Dias, A. F., Ramagem, S.*etal.* (2011). *Agricultural Research in Angola.* Challenges and proposals. Vol.1. 385 pp.

106. Sarandón, S., Zuluaga, M., Cieza, R., Gómez, C., Janjetic, L. and Negrete, E. (2006). *Evaluating the sustainability of farm systems in misiones, argentina, through the use of indicators. Agroecología*, 1, 19-28.

107. Sardinha, R. M. & Carriço, J. S. (1975). *Science, technology and the university in the development of the agricultural sector.* Reordering. Revista Junta Provincial de Povoamento de Angola, 38: 19-22.

108. Schuroff, I. A. (2005). *Cassava cultivation and family farming in the Lídia Ivinhema nucleus - MS. Lavras*, UFLA, 2005. 41p. (Specialization monograph).

109. Sebrae-Servicio de Apoio às Micro e pequenas Empresas de Bahia (2018).*Estudo de mercado. Agribusiness*: Horticulture. Brazil. Pp 45.

110. SEBRAE-SP. (2013). *Profile and needs of São Paulo's olive-growing* sector. Available at: http://m.sebrae.com.br/Sebrae/Portal%20Sebrae/UFs/SP/Pesquisas/relatorio_olericultura_paulista.pdf>. Accessed on: May 22, 2023.

111. Sousa, J. L and Resende, P. (2010). *Manual of organic horticulture.* Viçosa: UFV. Kristaponyte, I. (2015). Effect of fertilization system on the balance of planta nutrients and soil agrochemical proprieties. Agronomy Research. 3(1): pp 45 - pp 54.

112. Souza, A.M.K. (2018). *Growth and productivity of leaf cabbage in a slab system with different substrates* (Thesis presented to the Postgraduate Program in Agronomy). Center for Biological and Nature Sciences, Federal University of Acre. Rio Branco - AC.

113. Tello, S. Flavia Mabel (2009). *La participación política de las mujeresen los gobiernos locales latinoamericanos*: Barreras y desafíos para una efectiva democracia de género. Barcelona.

114. Tolio, B. A. and Doula, Sheila Maria (2016). *Rural youth and institutional influences for staying in the countryside: a case study in an agricultural cooperative in the Triângulo Mineiro INTERAÇÕES*, Campo Grande, MS,v. 17, n. 3, p. 370-383, jul./set. 2016.

115. Trani, E. P., Tivelli, S. W. Blat, F., *et al.* (2015). *Leaf cabbage: from planting to post-harvest Campinas*: Instituto Agronômico, 2015. 36 p. online (Série Tecnologia Apta. Boletim Técnico IAC). ISSN 1809-7936.

116. Triani, P., Passos, F., Teodoro, M. C., Santos, V. J., Frare, P. (2015). *Liming and fertilization for cabbage cultivation.* Available at: http//www.iac.sp.gov.br|calagemguiabo.htm. Consultation date: April 3, 2023.

117. Vasques, F. J. M. (2010). *Agriculture in the Luinga region of Angola*: cultural aptitude and technological adaptations (Dissertation for a Master's Degree in Agronomic Engineering). Higher Institute of Agronomy, Technical University of Lisbon.

118. Vilar, M., Cartea, M. E., Padilla G., Soengas, P., Velasco, P. (2008). *Thepotential of kales as a promising vegetable crop.* Euphytica, v. 159, n. 1, p. 153-65.

119. Villaret, A. (2002). *El enfoque sistémico aplicado al análisis del medioagrícola*: introducción al marco teórico y conceptual. Praxis del desarrollo rural. Pradem/CICDA. Peru.

120. Victoria L. V, S (2023). *Evaluation of agro-productive indicators of the cultivation of leaf cabbage (Brassica oleracea var. manteiga) with the application of different sources of organic and chemical fertilization in the locality of Namacunde.* End-of-course work towards the Degree of Engineer. Option: Agronomy. Ondjiva Polytechnic Institute. Mandume Yandemufayo University. "Huila Cunene

121. Wikipedia (2009). *International Code of Botanical Nomenclature.* Available at: http://es.wikipedia.org/wiki/CodigoInternacionaldeNomenclaturaBotanic. Consultation date: April 25, 2021.

122. Wizniewsky, C. R. and Wizniewsky, J. G. (2009). *Rural development, family farming and territorialities.* Available at: http://www.google.com.br/url?sa=t&rct=j&q=escolaridade%20%20agricultura%

20familiar&source=web&cd=7&sqi=2&ved=0CFoQFjAG&url=http%3A%2F%2Fegal2009.easyplanners.info%2Farea06%2F6413_Wizniewsky_Carmen_Rejane.doc&ei=xz0uUfWkD6WD0QH-gIFA&usg=AFQjCNFCB2d0b9uNHweEajmxBo-XyU0e3g&cad=rja>.Accessed: March 17, 2023.

123. Yasmeen Kausar (2011). *Impact of educated farmer on Agricultural Product*; Journal of Public Administration and Governance, Vol. 1, No. 2. Available at: http://www.macrothink.org/journal/index. /948/pdf. Accessed on: March 2023.

124. Zamberlam, J. and Froncheti, A. (2001). *Ecological agriculture: preserving the small farmer and the environment*. Petrópolis: Vozes, 214p.

VIII. ANNEXES

8.1 Climate data

MAP OF MAXIMUM AND MINIMUM TEMPERATURES, RELATIVE HUMIDITY AND RAINFALL FOR THE FIRST QUARTER OF 2023 - CLIMATE DATA

Days	Maximum Tª °c			Minimum Tª °c			Relative humidity %			Precipitation mm		
	Jan	Feb	Marc	Jan	Feb	Marc	Jan	Feb	Marc	Jan	Feb	Marc
01	30,3	29,2	33,1	19,3	20,0	19,7	70	61	51			
02	30,4	30,9	29,8	21,0	21,5	19,1	61	51	66			
03	23,7	30,1	32,8	19,8	21,3	19,8	67	67	44			
04	20,1	32,1	33,0	18,7	20,8	18,0	98	48	20			
05	29,8	31,4	34,0	19,0	22,2	18,4	40	30	22			
06	31,1	30,9	35,9	19,6	20,8	18,8	29	59	21			
07	33,7	26,7	35,9	20,8	20,9	21,4	31	58	35			
08	32,3	28,6	31,0	22,9	21,7	21,5	27	65	50			
09	31,3	31,7	34,7	21,2	19,4	22,6	30	43	30			
10	33,2	33,1	34,4	23,1	18,8	22,4	44	46	40			
11	28,9	33,1	32,0	21,2	20,0	20,8	75	31	78			
12	31,6	33,8	32,8	21,6	20,3	22,6	67	31	68			
13	30,7	35,1	34,8	19,9	21,7	21,0	62	35	40			
14	31,1	35,6	35,0	18,0	23,4	21,2	68	31	39			
15	31,9	35,2	34,3	18,0	21,6	21,1	50	22	33			
16	29,6	33,3	33,4	20,7	18,2	20,2	71	27	47			
17	26,4	32,9	31,7	20,0	16,9	17,7	67	31	40			
18	25,7	31,9	33,5	19,0	13,7	20,0	85	23	50			
19	25,1	33,0	33,3	18,1	15,0	19,4	96	22	46			
20	22,3	33,1	33,9	20,6	15,4	19,3	90	24	49			
21	27,2	34,2	32,6	18,6	14,8	20,2	78	22	55			
22	24,6	35,4	32,4	17,2	17,4	21,3	71	26	47			
23	27,4	34,0	29,8	19,4	21,3	18,8	79	39	72			
24	28,4	32,9	32,0	18,5	20,8	20,5	79	48	53			
25	30,2	34,5	30,8	20,6	20,5	18,5	68	49	46			
26	31,0	34,3	25,6	19,0	20,2	21,2	71	46	84			
27	31,4	32,5	30,6	20,2	21,6	18,3	70	64	55			
28	30,1	27,9	28,4	21,1	19,7	18,6	68	-	77			
29	30,3		31,0	20,1		20,3	59		66			
30	27,8		31,8	17,9		20,6	77		63			
31	27,4			19,6		18,9	76		62			
Total	**901,6**	**906,3**	**1006,6**	**548,8**	**548,8**	**622,2**		**1099,0**	**1565,0**			

Aver age	28.6	30,2	32,5	19,7	19,6	20,1	65	78	51			

8.2 Cabbage production area (*Brassica Olerácea* L. *var. troncuda o português* and *var. manteiga* I -1811 *(B)*) at the R. L Cantinho do Chicote Cooperative (N.J.F), in Mukolongodjo commune, Cuvelai municipality.

A. Evidence of the initial preparation of the nurseries.

B. Nurseries in development

C. Start of transplant

D. Plot in production

Printed by Books on Demand GmbH, Norderstedt / Germany